VAIROS

Data Driven Strategy

EXECUTIVE GAME PLAN

J. Tod Fetherling

J. Tod Fetherling

Publisher

7011 Ellendale Drive

Brentwood, TN. 37027

ISBN: 979-8-9928531-6-2

February 2026

CONTENTS

INTRODUCTION

The Vairos Planning System represents a disciplined approach to strategic planning built on the principle of Data Driven Strategy. In today's business environment, organizations are inundated with data yet find themselves with less time to focus on meaningful execution. This Executive Game Plan provides CEOs and senior leaders with a clear, actionable framework to transform data into a strategic advantage.

The essence of Vairos is victory in the right and opportune moment. Obtaining victory requires a well-crafted strategy. Strategy is not merely a document or an annual exercise; it is the continuous process of assessing the situation, making informed decisions, executing with discipline, and measuring success against objectives.

This condensed guide distills the Vairos methodology into its essential components, providing executives with the high-level understanding necessary to lead strategic planning initiatives within their organizations. The framework operates through three interconnected phases: Assess, Decide, and Execute.

As a CEO, your role is not to perform every analytical task or execute every tactical initiative. Your role is to ensure that your organization has the right framework, the right people, and the right systems in place to transform strategic intent into measurable results. This guide will equip you with that capability.

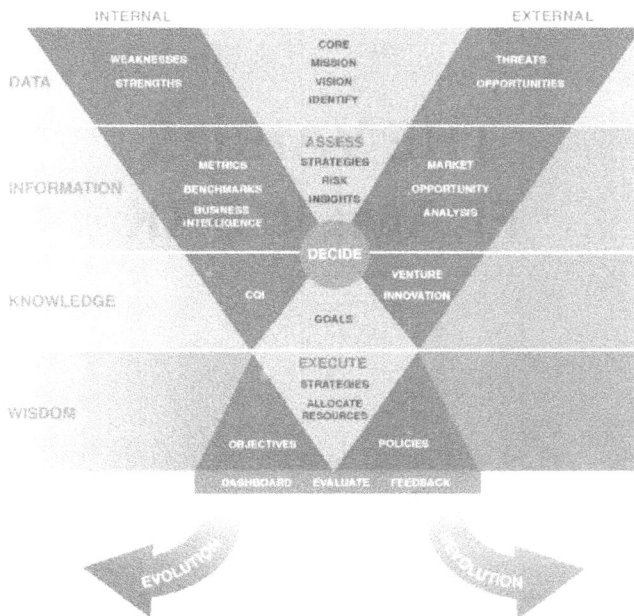

DATA DRIVEN STRATEGIC PLANNING

INTERNAL — EXTERNAL

DATA: WEAKNESSES, STRENGTHS, CORE, MISSION, VISION, IDENTIFY, THREATS, OPPORTUNITIES

INFORMATION: METRICS, BENCHMARKS, BUSINESS INTELLIGENCE, ASSESS, STRATEGIES, RISK, INSIGHTS, MARKET, OPPORTUNITY, ANALYSIS

KNOWLEDGE: DECIDE, CQI, VENTURE, INNOVATION, GOALS

WISDOM: EXECUTE, STRATEGIES, ALLOCATE, RESOURCES, OBJECTIVES, POLICIES, DASHBOARD, EVALUATE, FEEDBACK

EVOLUTION — EVOLUTION

THE VAIROS
FRAMEWORK

The Vairos Planning System translates the data continuum into a comprehensive strategic planning methodology. The framework operates through three major components: Assess, Decide, and Execute. Each component builds upon the previous, creating a systematic approach to data-driven strategy.

The Three Pillars

- **ASSESS** forms the foundation of all strategic work. Here, organizations conduct a comprehensive analysis of their current position, examining both internal capabilities and external market conditions. This phase ensures that all subsequent decisions remain grounded in reality rather than assumptions.

- **DECIDE** represents the critical juncture at which analysis transforms into a strategic choice. This is where the complexity of assessment narrows into a focused strategic direction. The Decide phase requires both analytical rigor and leadership judgment to select the most promising strategic paths.

- **EXECUTE** translates strategic decisions into measurable action. This phase breaks broad strategic direction into specific goals, objectives, strategies, tactics, and policies. It establishes measurement systems to track progress and ensure accountability.

The Strategic Data Continuum

From extensive research and experience over 1,000s of planning sessions, a Strategic Data Continuum emerged comprising six stages:

- **Stage 1 - Discovery:** What do we have? What does it tell us? (Descriptive Statistics)

- **Stage 2 - Integration:** Market data and external systems (Diagnostic Statistics)

- **Stage 3 - Business Intelligence:** Internal business system reporting

- **Stage 4 - Predictive Analytics:** Perform, Test, Replicate (Agile Analytics)

- **Stage 5 - Execution:** Turning Data into Action (Prescriptive Analytics)

- **Stage 6 - Decision Sciences:** Capturing decisions based on available data and feeding back into Stage 1

The Data-Driven Difference

What distinguishes Vairos from traditional strategic planning approaches is its systematic integration of data throughout all three phases. Rather than relying solely on intuition or experience, Vairos demands evidence-based analysis at each stage.

This data-driven approach provides several advantages: Objectivity that reduces bias in both assessment and decision-making; Accountability through measurable standards for success; Agility that enables rapid course correction based on performance data; and continuous

Learning that builds organizational capability for improvement.

Implementation Principles

The Vairos system operates on several core principles that guide implementation:

- **Systematic Approach:** Each phase builds logically on the previous, ensuring comprehensive coverage without overwhelming complexity.

- **Scalable Framework:** The system works for organizations of all sizes, from startups to established enterprises, adapting to organizational context while maintaining methodological rigor.

- **Iterative Process:** While the phases appear linear, Vairos operates as a continuous cycle, with insights from execution feeding back into future assessments.

- **Collaborative Methodology:** The system encourages cross-functional participation while maintaining clear accountability and decision-making authority.

ASSESS: THE FOUNDATION

Before any organization can develop an effective strategy, it must understand its fundamental identity. The ASSESS component of Vairos focuses on establishing this foundation: the essential elements that define who you are, what you stand for, and how you operate.

Most businesses operate in reactive mode, responding to immediate pressures without clear strategic direction. Vairos transforms organizations into proactive, highly efficient, highly effective entities by first establishing a solid core foundation.

The Foundation Metaphor

Consider working with a personal trainer. The first step is not selecting exercises or setting fitness goals; it is understanding motivation and building core strength. Without core strength, advanced movements become dangerous and ineffective. The same principle applies to organizational strategy.

Once you establish your organizational core, you can move to more advanced strategic initiatives. However, skipping this foundational work, even if you believe you already understand your organization's core, typically leads to strategic stumbling later in the process.

Assess Components

The Assess phase encompasses five critical elements:

1. **Mission:** The inspirational purpose that drives daily operations and speaks to the heart of why your organization exists.

2. **Vision:** The aspirational future state that guides strategic direction and appeals to the imagination of what is possible.

3. **Identity:** Understanding the key attributes and characteristics that define your organization's unique nature.

4. **Strategic Review:** Systematic examination of past strategic efforts to understand what works and what does not.

5. **Risk Assessment:** Honest evaluation of potential threats and vulnerabilities that could derail strategic initiatives.

Business Model Clarity

Before proceeding with the core definition, you must clearly understand two fundamental concepts:

- **Business Model:** HOW does the business work? This requires a detailed understanding of production processes, revenue drivers, and essential operating elements. Are you primarily a human capital organization, a technology-driven entity, or a manufacturing concern? Understanding your operating model influences every strategic decision.

- **Business Plan:** How does your organization turn ideas into revenue? This encompasses everything from market analysis to financial projections, from competitive positioning to growth strategies.

The Six Operating Models

Organizations typically operate under one of six primary models, each requiring different strategic approaches:

1. **Partnership Model:** Highly reliant on people and expertise, common in professional services. Success depends on advancing through partnership ranks and consistently generating a new stream of partners to fuel growth.

2. **Matrix Model:** Requires thinking about multiple sides of the business simultaneously, with incentives aligned across different organizational dimensions.

3. **Diversification Model:** Spreads risk across multiple industries, products, or services, requiring sophisticated portfolio management.

4. **Assembly Line Model:** Emphasizes communication and collaboration among departments, often requiring technology or manufacturing processes for coordination.

5. **Parts = Sum of Whole Model:** Treats each business unit as an independent profit center, requiring strong individual accountability.

6. **Unity Model:** Focuses on single service or product excellence, emphasizing efficiency and brand development.

Stakeholder Analysis

Every organization, regardless of size, has key stakeholders whose interests that you must consider in strategic planning. These typically include customers, employees, shareholders, partners, and community members.

Beyond identifying stakeholders, you must understand their relative importance and influence on organizational success. This ranking provides guidance for strategic decision-making and resource allocation.

MISSION

Mission represents the heart of your organization, the inspirational purpose that drives daily operations and connects every employee to meaningful work. While vision appeals to the imagination, mission speaks directly to the heart, providing emotional energy for day-to-day activities.

The Heart Connection

Mission statements must evoke intense passion. They should be memorable enough to be embedded in every team member's consciousness. Your organizational mission should provide both inspiration and practical guidance for decision-making.

Characteristics of Effective Missions

- **Inspirational Rather Than Aspirational:** Mission differs from vision in its immediate relevance to current operations. While vision describes future possibilities, mission explains current purpose.

- **Heart-Centered:** Mission statements must connect emotionally with team members, customers, and stakeholders. They should explain why the work matters.

- **Action-Oriented:** Effective missions describe what the organization does and why it matters, not just what it hopes to become.

- **Culturally Aligned:** The mission must reflect the values and behaviors that drive organizational success.

Mission Development Rules

- **Avoid Ambiguous Language:** Every word must have a clear, specific meaning without the risk of misinterpretation.

- **Use Simple Structure:** Avoid long, complex sentences. Your mission statements should be memorable and repeatable.

- **Choose Accessible Vocabulary:** Unless your organization requires technical language, use words that any team member can understand and embrace.

- **Keep It Simple:** The best missions can be "tattooed on the hearts" of employees. Complexity destroys memorability.

- **Create Emotional Connection:** Mission must generate passion and commitment, not just intellectual understanding.

- **Ensure Alignment:** Mission should guide behavior and decision-making at all organizational levels.

Mission as Decision Filter

An effective mission statement serves as a filter for decision-making. When facing strategic choices, ask: "Does this option advance our mission?" If the answer is unclear or negative, reconsider the decision.

This filtering function transforms the mission from wall decoration into a practical management tool. Teams can use a mission to evaluate projects, allocate resources, and resolve conflicts.

VISION

Vision represents the aspirational future that captures the imagination and drives strategic ambition. While the mission speaks to the heart about the current purpose, vision engages the mind about future possibilities. Vision describes the "what", including what the organization is building toward and what success will look like.

The Brain Connection

Vision statements need to evoke emotions that appeal to the dreamer, the strategist, the part of organizational consciousness that imagines what could be rather than what is. Effective vision provides direction for long-term strategic planning while inspiring commitment to difficult transformation work.

Characteristics of Effective Visions

- **Aspirational and Specific:** Vision should stretch organizational imagination while remaining concrete enough to guide decisions.

- **Future-Focused:** Vision describes tomorrow's organization, not today's reality.

- **Emotionally Compelling:** Like the Mission, the Vision must connect emotionally, but with hope and excitement rather than with the current purpose.

- **Strategically Relevant:** Vision should inform resource allocation and strategic choice, not just provide inspirational language.

Vision Development Rules

- **Be Short and Memorable:** Every team member must be able to remember and repeat the vision.

- **Be Organizationally Specific:** Vision should apply uniquely to your organization, not to any company in your industry.

- **Be Purpose-Driven:** Vision should explain why the future state matters, not just what it includes.

- **Align with Culture and Values:** Vision must reflect authentic organizational character, not aspirational culture that does not exist.

- **Provide Strategic Guidance:** Vision should inform major decisions about resource allocation and strategic direction.

Vision and Mission Relationship

Vision and mission work together to provide complete organizational guidance:

- Mission explains the current purpose and daily motivation.

- Vision describes future aspirations and strategic direction.

Both must be authentic to organizational reality and inspiring to organizational stakeholders. Both should influence decision-making and resource allocation.

IDENTIFY

The Identify component focuses on understanding the human dynamics that will determine strategic success or failure. This involves mapping key stakeholders, understanding organizational influencers, and recognizing the informal networks that drive implementation.

The Human Factor in Strategy

Technical strategies fail when they ignore human realities. The most sophisticated plans become worthless without people who can and will execute them. Identification work ensures that strategy accounts for organizational politics, informal influence networks, and individual motivations that drive behavior.

Stakeholder Mapping

Begin with systematic identification of key organizational stakeholders:

- **Internal Stakeholders:** Employees, managers, executives, board members, and union representatives.

- **External Stakeholders:** Customers, suppliers, partners, community leaders, regulatory bodies, and investors.

- **Influential Networks:** Industry associations, professional groups, informal advisory relationships.

For each stakeholder group, document:

- ⋏ Level of influence on strategic outcomes

- ⋏ Current attitude toward the change curve and initiatives

- ⋏ Potential contribution to strategy execution

- ⋏ Specific concerns or interests that must be addressed.

The "Doers" Analysis

Every organization has people who get stuff done (#GSD), the informal leaders who can accelerate or derail strategic initiatives. These individuals may lack formal authority but command respect, relationships, and practical knowledge that drive real change. Ask these diagnostic questions:

- ⅄ When the organization faced a major challenge, who stepped up?
- ⅄ Who organized the response to the last crisis?
- ⅄ If you needed something done quickly, who would you call?
- ⅄ Who was missing when things went wrong?

The answers reveal your organization's informal power structure, including the information crucial for implementation planning.

Change Champions and Resisters

Strategic change elicits predictable human responses. Some individuals embrace change as an opportunity; others resist it as a threat. A successful strategy requires understanding both camps.

- **Change Champions:** They typically demonstrate comfort with ambiguity; Show a track record of adaptation; Possess influence with key stakeholder groups; Express dissatisfaction with the current state.

- **Change Resisters:** They typically benefit from current systems and processes; Fear loss of status, authority, or job security; Lack confidence in the organization's change capability; Have been burned by previous failed initiatives.

Neither group is inherently good or bad. Both responses are rational given individual circumstances. An effective strategy accounts for both and creates approaches that address legitimate concerns while building momentum for change.

REVIEW STRATEGIES

Before developing new strategic directions, organizations must systematically examine their strategic history. The Review Strategies component evaluates past strategic initiatives to understand what worked, what failed, and why. This analysis provides crucial intelligence for future strategic decision-making.

Learning from Strategic History

Most organizations repeat strategic mistakes because they fail to systematically analyze their strategic track record. The Review Strategies component creates organizational memory about strategic effectiveness, helping leaders make more informed decisions about future initiatives.

This review examines multiple dimensions of past strategic performance:

- **Outcomes:** Did strategies achieve intended results?

- **Process:** How well did planning and implementation work?

- **Context:** What environmental factors influenced the results?

- **Learning:** What insights can inform future strategy?

Strategic Initiative Inventory

Begin by cataloging strategic initiatives from the past three to five years:

- **Major Strategic Plans:** Comprehensive organizational strategies, five-year plans, and major transformational initiatives.

- **Significant Projects:** Large investments in technology, facilities, market expansion, and new service lines.

- **Operational Improvements:** Process redesign, efficiency initiatives, quality improvement programs.

- **Cultural Changes:** Leadership development, organizational restructuring, value implementation programs.

For each initiative, document:

- ⚲ Original objectives and success metrics;
- ⚲ Resources invested;
- ⚲ Actual outcomes and performance against goals;
- ⚲ Implementation challenges and unexpected obstacles;
- ⚲ Key decisions that influenced results.

Success Factor Analysis

Identify patterns in successful strategic initiatives:

- **Strategic Design:** What characteristics made strategies more likely to succeed?

    - ★ Clear, measurable objectives;
    - ★ Realistic resource requirements;
    - ★ Strong stakeholder alignment;
    - ★ Appropriate timing and sequencing.

- **Implementation Excellence:** What execution factors drove success?

  ★ Strong project management;

  ★ Effective communication;

  ★ Adequate resource allocation;

  ★ Consistent leadership support.

- **Environmental Conditions:** What external factors supported success?

  ★ Market timing;

  ★ Competitive dynamics;

  ★ Regulatory environment;

  ★ Economic conditions.

Failure Mode Analysis

Examine unsuccessful initiatives to understand failure patterns.

⅄ **Strategic Flaws:**

  ★ Unclear objectives or success metrics;

  ★ Unrealistic resource assumptions;

★ Poor stakeholder analysis;

★ Inadequate risk assessment.

▲ **Execution Problems:**

    ★ Insufficient project management;

    ★ Poor communication and change management;

    ★ Resource shortfalls or competing priorities;

    ★ Leadership turnover or loss of support.

ASSESS RISK

Risk assessment represents the critical reality check for all strategic thinking. While strategy development naturally focuses on opportunities and positive outcomes, effective planning requires honest evaluation of potential threats, vulnerabilities, and failures that could derail strategic initiatives.

Strategic Risk Categories

Strategic risks fall into multiple categories, each requiring different assessment and mitigation approaches:

- **Market Risks:** Changes in customer demand, competitive dynamics, or industry structure that could undermine strategic assumptions.

- **Operational Risks:** Internal capacity constraints, capability gaps, or execution failures that could prevent strategy implementation.

- **Financial Risks:** Resource shortfalls, cost overruns, or revenue disappointments that could force strategic modifications.

- **Regulatory Risks:** Policy changes, compliance requirements, or legal challenges that could block strategic initiatives.

- **Technology Risks:** System failures, cyber threats, or technological obsolescence that could disrupt operations.

- **Reputation Risks:** Brand damage, stakeholder conflicts, or public relations problems that could undermine strategic credibility.

Risk Assessment Methodology

Systematic risk assessment examines both probability and impact of potential problems:

- **Risk Identification:** Comprehensive catalog of potential threats to strategic objectives, developed through brainstorming, scenario planning, and historical analysis.

- **Probability Assessment:** Realistic evaluation of how likely each risk is to occur, based on historical data, expert judgment, and environmental analysis.

- **Impact Analysis:** Assessment of potential consequences if risks materialize, including financial costs, strategic delays, and organizational disruption.

- **Risk Prioritization:** Focus on the highest-priority risks based on a combination of probability and impact.

Scenario Development

Create multiple scenarios to test strategic resilience:

- **Best-Case Scenario:** If everything goes according to plan, what would success look like?

- **Most Likely Scenario:** Realistic expectations based on organizational history and market conditions.

- **Worst-Case Scenario:** Multiple problems occur simultaneously. How would the organization respond?

- **Black Swan Events:** Low-probability, high-impact events that could fundamentally alter the strategic context.

Risk Mitigation Planning

Develop specific approaches for managing high-priority risks:

- **Risk Avoidance:** Modifying strategic approaches to eliminate certain risks entirely.

- **Risk Mitigation:** Reducing the probability or impact of risks through preventive measures.

- **Risk Transfer:** Using insurance, partnerships, or contracts to shift risk to other parties.

- **Risk Acceptance:** Acknowledging risks that cannot be economically managed and planning for potential consequences.

- Contingency Planning: Developing specific response plans for high-impact risks.

DECIDE:
THE CRITICAL CHOICE

The Decide phase transforms strategic analysis into strategic action by making clear choices about organizational direction and priorities. Effective decision-making combines analytical rigor with leadership judgment to select strategies that advance organizational objectives while remaining realistic about implementation challenges and resource constraints.

Decision Science Integration

The Decide phase incorporates decision science principles to improve choice quality:

- **Data-Driven Analysis:** Using quantitative and qualitative data to inform decision-making.

- **Scenario Planning:** Testing strategic choices against multiple potential futures.

- Decision Tracking: Documenting decision rationale for future evaluation and learning.

- **Bias Recognition:** Understanding and managing cognitive biases that distort decision-making.

- Stakeholder Input: Gathering diverse perspectives while maintaining clear decision authority.

Behavioral Economics in Strategic Decisions

Strategic decisions involve human psychology and organizational behavior that influence outcomes:

- **Risk Tolerance:** Understanding organizational and leadership appetite for uncertainty and potential failure.

- **Loss Aversion:** Recognizing tendency to overweigh potential losses versus potential gains.

- **Anchoring Effects:** Managing influence of initial information or assumptions on subsequent analysis.

- **Confirmation Bias:** Actively seeking disconfirming evidence to test strategic assumptions.

- **Group Think:** Ensuring diverse perspectives and constructive dissent in decision processes.

Decision Process Design

Effective strategic decision-making requires structured processes:

- **Information Synthesis:** Converting assessment findings into decision-relevant insights.

- **Alternative Generation:** Creating multiple strategic options rather than evaluating single proposals.

- **Criteria Application:** Systematically evaluating alternatives against established decision criteria.

- Stakeholder Consultation: Gathering input from key stakeholders while maintaining decision authority.

- **Decision Documentation:** Recording decision rationale, assumptions, and expected outcomes.

LEADERSHIP DECISION-MAKING

Strategic decisions ultimately require leadership judgment that transcends analysis. The CEO's role is not to make every decision but to ensure that the organization has the right framework, the right people, and the right systems in place to make decisions well.

Vision Integration

Ensuring that decisions advance the organization's vision and mission. Every major decision should be tested against the question: Does this move us closer to our stated vision and fulfill our mission?

Stakeholder Balance

Managing competing interests and priorities among different stakeholder groups. Customers, employees, shareholders, and community members all have legitimate claims that must be balanced in strategic decision-making.

Resource Reality

Making decisions within realistic resource constraints. Ambition without resources is merely hope. Effective decisions account for the financial, human, and technological resources available or obtainable.

Implementation Capability

Choosing strategies that align with organizational execution capabilities. The best strategy is worthless if the organization lacks the capability to implement it effectively.

Change Management

Considering organizational capacity for strategic change and transformation. Even brilliant strategies fail when they exceed the organization's ability to absorb and execute change.

Decision Timing and Sequencing

Strategic decisions involve critical timing considerations:

- **Market Timing:** Aligning decisions with market conditions and competitive dynamics.

- **Organizational Readiness:** Ensuring organizational capability to implement decisions effectively.

- **Resource Availability:** Coordinating decisions with resource availability and budget cycles.

- **Dependency Management:** Sequencing decisions to manage interdependencies among strategic initiatives.

- **Window of Opportunity:** Acting within time constraints imposed by external conditions.

Decision Validation and Testing

Strategic decisions benefit from validation before full implementation:

- **Pilot Programs:** Testing strategic approaches on limited scale before major resource commitments.

- **Stakeholder Feedback:** Validating decisions with customers, employees, and other key stakeholders.

- **Competitive Analysis:** Ensuring decisions account for likely competitive responses.

- **Financial Modeling:** Testing economic assumptions underlying strategic decisions.

- **Risk Assessment:** Evaluating potential negative consequences and mitigation strategies.

FROM DECISION TO EXECUTION PLANNING

Strategic decisions must translate into execution plans. The gap between strategic intent and operational reality is where most strategies fail. This section addresses how to bridge that gap effectively.

Goal Setting

Converting strategic decisions into specific, measurable objectives. Goals should be SMART: Specific, Measurable, Achievable, Relevant, and Time-bound. Each goal should clearly connect to the strategic decision it implements.

Resource Allocation

Distributing financial, human, and technological resources to support decisions. Resource allocation is where strategy becomes real, where priorities become tangible through investment. A strategy without resources is merely a wish.

Timeline Development

Creating realistic schedules for strategic implementation. Timelines must account for dependencies, organizational capacity, and market conditions. Unrealistic timelines destroy credibility and momentum.

Accountability Assignment

Designating responsibility for strategic execution. Every goal needs a name next to it, not a committee, not a task force, but a person who is responsible and accountable for results.

Performance Measurement

Establishing metrics to track progress and success. What gets measured gets managed. Effective metrics include both leading indicators that predict future success and lagging indicators that confirm results.

Decision Communication

Strategic decisions require effective communication throughout the organization:

- **Rational Explanation:** Helping stakeholders understand the reasoning behind strategic choices.

- **Expectation Setting:** Clearly communicating what decisions will and will not accomplish.

- **Role Clarification:** Explaining how decisions affect different organizational functions and individuals.

- **Change Management:** Preparing the organization for changes resulting from strategic decisions.

- **Feedback Mechanisms:** Creating channels for ongoing input and course correction.

Common Decision-Making Failures

- **Analysis Paralysis:** Gathering excessive information without making necessary decisions.

- **Premature Closure:** Making decisions without adequate analysis or consideration of alternatives.

- **Resource Optimism:** Underestimating resource requirements for strategic implementation.

- Complexity Underestimation: Failing to account for implementation challenges and dependencies.

- **Stakeholder Misalignment:** Making decisions without adequate stakeholder buy-in or support.

EXECUTE: FROM STRATEGY TO RESULTS

The Execute phase translates strategic decisions into measurable action through systematic implementation planning and performance management. This phase represents the bridge between strategic intention and operational reality, ensuring that strategic choices drive actual organizational behavior and results.

From Decision to Discipline

Execution transforms abstract strategic concepts into concrete organizational activities. This transformation requires breaking broad strategic directions into specific components that you can assign, measure, and manage effectively.

The Execution Architecture consists of:

- **Goals:** Strategic objectives that define desired outcomes;

- **Objectives:** Specific milestones that indicate progress toward goals;

- **Strategies:** Approaches for achieving objectives and goals;

- **Tactics:** Specific initiatives and activities that implement strategies;

- **Policies:** Guidelines and rules that govern execution behavior.

Strategic Focus and Prioritization

Most organizations can effectively execute only three to five core strategies simultaneously. Beyond this limit, organizational bandwidth becomes strained, resulting in poor execution across all initiatives.

Focus Principles:

- **Selective Excellence:** Choosing fewer initiatives and executing them exceptionally well;

- **Resource Concentration:** Allocating adequate resources for competitive advantage;

- **Capability Alignment:** Matching strategic initiatives with organizational strengths;

- **Sequential Execution:** Timing strategic initiatives to manage organizational capacity.

The Execution Hierarchy

Effective execution requires clear hierarchy that connects strategic vision with daily action:

- **Strategy Level:** Broad direction and approach for achieving competitive advantage.

- **Goal Level:** Measurable outcomes that indicate strategic progress.

- **Objective Level:** Specific milestones and deliverables within defined timeframes.

- **Tactic Level:** Individual projects, initiatives, and activities.

- **Policy Level:** Guidelines that govern how execution occurs.

Strategic vs. Operational Goals

Strategic goals focus on transformation and competitive advantage, while operational goals maintain current performance levels:

- **Strategic Goals:** Market expansion, new capability development, competitive repositioning.

- **Operational Goals:** Efficiency improvement, quality maintenance, cost management.

Both types are necessary, but strategic planning should emphasize strategic goals that advance organizational positioning rather than simply maintaining current operations.

GOALS

Goals represent the strategic targets that define what success looks like for major organizational initiatives. In the Vairos framework, goals serve as the North Star for execution, providing clarity, alignment, and accountability for strategic progress.

The Strategic Role of Goals

Goals bridge the gap between strategic vision and operational action. They translate abstract strategic concepts into concrete targets that guide resource allocation, performance measurement, and organizational behavior.

Goal Characteristics:

- ⅄ Strategic Relevance, directly connected to strategic direction and competitive advantage;

- ⅄ Measurable Outcomes, Quantifiable results that indicate success or failure;

- ⅄ Time-Bound Achievement, Specific deadlines that create urgency and accountability;

- ⅄ Motivational Power, Ambitious enough to inspire effort while remaining achievable.

SMART Goal Framework

- **Specific:** Goals must clearly define what will be accomplished, avoiding vague or ambiguous language.

- **Measurable:** Goals require quantifiable indicators that demonstrate progress and success.

- **Achievable:** Goals should stretch organizational capability while remaining within realistic bounds.

- **Relevant:** Goals must connect meaningfully to strategic objectives and stakeholder value.

- **Time-Bound:** Goals need specific deadlines that create urgency and enable accountability.

Goal Categorization and Portfolio

- **Financial Goals:** Revenue growth, profit improvement, cost reduction, return on investment.

- **Market Goals:** Market share expansion, customer acquisition, geographic expansion, competitive positioning.

- **Operational Goals:** Efficiency improvement, quality enhancement, capacity development, process optimization.

- **Innovation Goals:** New product development, technology adoption, capability building, research advancement.

- **Stakeholder Goals:** Customer satisfaction, employee engagement, community impact, regulatory compliance.

Goal Ownership and Accountability

- **Executive Sponsorship:** Senior leadership ownership for strategic goal achievement.

- **Functional Responsibility:** Department or team accountability for specific goal components.

- **Individual Accountability:** Personal responsibility for goal-related activities and outcomes.

- **Cross-Functional Coordination:** Collaboration among different organizational units for goal achievement.

- **Performance Consequences:** Clear linkage between goal achievement and organizational rewards.

OBJECTIVES

Objectives represent the specific milestones and deliverables that indicate progress toward strategic goals. While goals define ultimate success, objectives break the journey into manageable components that maintain momentum and enable course correction throughout strategic execution.

Objectives as Strategic Milestones

Objectives answer the question: "How will we know we are making progress?" They provide intermediate targets that build confidence, maintain motivation, and enable adaptive management of strategic initiatives.

Objective Characteristics:

- ⅄ Milestone Nature, Specific achievements that mark progress toward larger goals;

- ⅄ Time-Bound Delivery, Clear deadlines that create accountability and urgency;

- ⅄ Measurable Results, Quantifiable outcomes that indicate completion;

- ⅄ Sequential Logic, Logical progression that builds toward goal achievement.

The Goal-Objective Relationship

Goals define what success looks like at the strategic level. Objectives specify how progress toward goals will be achieved and measured.

Example:

- **Goal:** Achieve 10% market share in cardiovascular services in Region North by Q4 next year.

- **Objectives:**

  - ★ Open two new cardiology clinics in Region North by Q2;

- ★ Recruit and onboard three new cardiologists by end of Q3;

- ★ Establish referral partnerships with ten primary care practices by mid-year;

- ★ Achieve 50 cardiovascular procedures per month by Q3.

Objective Development Process

- • **Goal Decomposition:** Breaking strategic goals into component achievements required for success.

- • **Timeline Mapping:** Sequencing objectives to create logical progression and manage dependencies.

- • **Resource Assessment:** Ensuring adequate resources are available for objective achievement.

- • **Dependency Analysis:** Understanding how objectives relate to and depend upon each other.

- • **Risk Evaluation:** Identifying potential obstacles to objective achievement and mitigation strategies.

Objective Ownership and Accountability

- **Clear Assignment:** Specific individuals responsible for objective achievement.

- **Authority Alignment:** Ensuring objective owners have authority to drive completion.

- **Resource Access:** Providing objective owners with necessary resources and support.

- **Performance Linkage:** Connecting objective achievement with individual and team performance evaluation.

- **Escalation Processes:** Clear procedures for addressing objective difficulties or delays.

STRATEGIES

Strategies represent the fundamental approaches that organizations use to achieve competitive advantage and strategic objectives. In the Vairos framework, strategies define how organizations will compete, create value, and position themselves for long-term success.

Strategic Intent and Direction

Strategies answer three critical questions that define organizational direction:

- **Where Will We Play?** Markets, customer segments, geographic regions, and competitive arenas where the organization will focus its efforts.

- **How Will We Win?** Competitive advantages, differentiation approaches, and value propositions that will drive success in chosen markets.

- **What Capabilities Must We Build?** People, technologies, processes, and resources are required to execute successfully and maintain a competitive advantage.

Core Strategy Types

- **Differentiation Strategy:** Creating unique value propositions that distinguish the organization from competitors through superior quality, innovation, service, or customer experience.

- **Cost Leadership Strategy:** Achieving competitive advantage through superior operational efficiency, enabling competitive pricing while maintaining acceptable margins.

- **Focus Strategy:** Concentrating resources on specific market segments, customer needs, or geographic areas to achieve deep specialization and competitive advantage.

- **Innovation Strategy:** Building competitive advantage through continuous development of new products, services, technologies, or business models.

- **Partnership Strategy:** Leveraging strategic alliances, joint ventures, and ecosystem relationships to access capabilities and markets beyond organizational boundaries.

Strategic Coherence and Integration

Effective strategies create coherence across organizational activities:

- **Resource Alignment:** Financial, human, and technological investments support strategic direction.

- **Capability Integration:** Organizational skills and competencies reinforce strategic positioning.

- **Cultural Consistency:** Values, behaviors, and practices support strategic execution.

- **System Coordination:** Policies, processes, and structures enable strategic implementation.

- **Stakeholder Alignment:** Customer, employee, and partner relationships advance strategic objectives.

Common Strategic Mistakes

- **Strategy Proliferation:** Attempting to pursue too many strategies simultaneously.

- **Generic Positioning:** Choosing strategies that do not create meaningful differentiation.

- **Capability Mismatch:** Selecting strategies that exceed organizational capabilities.

- **Market Misunderstanding:** Developing strategies based on inaccurate market assumptions.

- **Implementation Neglect:** Focusing on strategy development while ignoring execution requirements.

TACTICS

Tactics represent the specific initiatives, projects, campaigns, and actions that implement strategic approaches and achieve strategic objectives. While strategies define broad direction, tactics specify the detailed work that transforms strategic intentions into organizational reality.

Tactics as Strategic Implementation

Tactics bridge the gap between strategic planning and operational execution by translating abstract strategic concepts into concrete activities that can be assigned, scheduled, and measured.

Tactical Characteristics:

- **Specific Actions:** Detailed initiatives with clear scope and deliverables;

- **Assigned Responsibility:** Individual or team ownership for tactical execution;

- **Defined Timelines:** Specific start and completion dates for tactical activities;

- **Resource Requirements:** Clear understanding of financial, human, and technological needs;

- **Measurable Outcomes:** Quantifiable results that indicate tactical success.

The Strategic Execution Stack

- **Strategy:** Broad approach for achieving competitive advantage.

- **Goal:** Measurable outcome that indicates strategic progress.

- **Objective:** Specific milestone towards goal achievement.

- **Tactic:** Individual initiative that advances objectives.

Example:

- ⋏ **Strategy:** Expand cardiovascular services in Region North.

- ⋏ **Goal:** Achieve 10% market share by Q4 next year.

- ⋏ **Objective:** Open two new clinics by Q2.

- ⋏ **Tactic:** Launch facility buildout project and hire project manager by end of month.

Tactical Categories and Types

- **Market Development Tactics:** Activities that build market presence and customer relationships, Customer acquisition campaigns; Market research and analysis; Brand development initiatives; Distribution channel development.

- **Capability Building Tactics:** Initiatives that develop organizational capabilities, Talent recruitment and development; Technology implementation projects; Process improvement initiatives; Training and development programs.

- **Operational Excellence Tactics:** Activities that improve operational effectiveness, Efficiency improvement projects; Quality enhancement initiatives; Cost reduction programs; Performance optimization efforts.

- **Innovation Tactics:** Initiatives that develop new capabilities and opportunities, Research and development projects; Product development initiatives; Technology pilot programs; Partnership development efforts.

Tactical Success Factors

- **Clear Definition:** Detail tactics specifically, with unambiguous scope and deliverables.

- **Adequate Resources:** Sufficient financial, human, and technological support for tactical success.

- **Strong Ownership:** Committed individuals with capability and authority to drive tactical completion.

- **Regular Monitoring:** Consistent attention to tactical progress and obstacle resolution.

- **Strategic Connection:** Clear linkage between tactical activities and strategic objectives.

POLICIES

Policies represent the strategic guardrails that enable autonomous execution while maintaining organizational coherence and compliance. In the Vairos framework, policies are not bureaucratic constraints but strategic enablers that provide clarity and consistency for decision-making throughout the organization.

Policy as Strategic Infrastructure

Policies answer the fundamental question: "What rules govern how we operate?" In a data-driven, decentralized environment where execution often occurs far from strategic leadership, policies become essential for maintaining strategic alignment and operational effectiveness.

Policy Functions:

- **Reduce Ambiguity:** Clear expectations enable faster decision-making and execution;

- **Ensure Compliance:** Consistent adherence to legal, regulatory, and ethical requirements;

- **Protect Brand:** Unified standards maintain organizational reputation and stakeholder confidence;

- **Enable Scale:** Systematic rules allow growth without proportional management overhead;

- **Support Strategy:** Operational guidelines that reinforce strategic direction and priorities.

Policy Categories for Strategic Execution

Strategic Policies:

- Guidelines that govern strategic decision-making and resource allocation

- Investment criteria and approval processes

- Market entry and partnership standards

- Innovation and R&D guidelines

- Competitive response protocols.

Operational Policies:

- Rules that ensure consistent operational execution
- Quality standards and procedures
- Customer service requirements
- Supplier and vendor management
- Performance management criteria.

Compliance Policies:

- Requirements that ensure legal and regulatory adherence
- Data privacy and security standards
- Financial reporting and controls
- Safety and environmental requirements
- Professional and ethical standards.

Governance Policies:

- Frameworks that guide organizational decision-making
- Authority and approval levels

- Communication and reporting requirements
- Risk management procedures
- Change management processes.

Policy Success Factors

- **Strategic Relevance:** Policies that clearly support organizational objectives and strategic direction.

- **Practical Application:** Policies that can be realistically implemented and enforced.

- **Clear Communication:** Policies that are well-understood throughout the organization.

- **Regular Review:** Systematic processes for policy evaluation and improvement.

- **Cultural Integration:** Policies that reinforce desired organizational culture and behaviors.

MEASURE

The Measure component transforms strategic execution into quantifiable insights through systematic performance tracking and analysis. Measurement provides the feedback loop that enables learning, accountability, and continuous improvement throughout the strategic planning cycle.

Strategic Measurement Philosophy

In the Vairos framework, measurement serves multiple critical functions beyond simple performance tracking:

- **Progress Validation:** Confirming that strategic initiatives are advancing according to plan.

- **Course Correction:** Providing early warning of problems that require adaptive management.

- **Accountability:** Creating objective standards for evaluating individual and organizational performance.

- **Learning:** Generating insights that improve future strategic planning and execution.

- **Stakeholder Communication:** Demonstrating progress and value to customers, investors, and other stakeholders.

Measurement Architecture

Effective strategic measurement requires integrated systems that capture multiple dimensions of performance:

- **Dashboard Systems:** Visual representations of strategic progress across all major initiatives.

- **Evaluation Processes:** Systematic assessment of strategic effectiveness and competitive impact.

- **Feedback Mechanisms:** Structured approaches for capturing and applying lessons learned from strategic execution.

The Measurement Hierarchy

- Strategic Level: Overall progress toward strategic objectives and competitive positioning.

- **Goal Level:** Achievement of specific strategic goals and major milestones.

- **Objective Level:** Completion of intermediate objectives and tactical deliverables.

- **Activity Level:** Execution of specific initiatives and operational activities.

- **Leading vs. Lagging Indicators:** Combination of predictive metrics and outcome measures.

Measurement Integration with Strategy

- **Strategy Development:** Using measurement insights to inform strategic choices and priorities.

- **Resource Allocation:** Applying performance data to guide investment decisions.

- **Implementation Management:** Tracking execution progress and identifying obstacles.

- **Strategic Adaptation:** Using measurement feedback to refine and improve strategic approaches.

- **Stakeholder Engagement:** Communicating strategic progress and value to key constituencies.

DASHBOARD

Strategic dashboards provide real-time visualization of organizational performance across all major strategic initiatives. In the Vairos framework, dashboards serve as the strategic cockpit that enables leaders to monitor progress, identify issues, and make informed decisions about resource allocation and course correction.

Dashboard as Strategic Cockpit

Dashboards transform complex performance data into accessible visual information that supports strategic decision-making:

- **Real-Time Visibility:** Status of strategic initiatives and performance trends.

- **Exception Management:** Highlighting areas that require leadership attention or intervention.

- **Pattern Recognition:** Identifying trends and relationships across different performance dimensions.

- **Decision Support:** Providing information needed for strategic choices and resource allocation.

- **Accountability:** Creating transparency about performance and responsibility.

Dashboard Design Principles

- **Role-Based Information:** Different views for executives, managers, and individual contributors.

- **Strategic Focus:** Emphasis on metrics that directly relate to strategic objectives and competitive advantage.

- **Visual Clarity:** Clear, intuitive presentation that enables rapid understanding and decision-making.

- **Actionable Insights:** Information that leads to specific actions rather than passive observation.

- **Balanced Perspective:** Combination of financial, operational, customer, and innovation metrics.

Executive Dashboard Components

- **Strategic Goal Progress:** Visual representation of advancement toward major strategic objectives.

- **Key Performance Indicators:** Critical metrics that indicate overall organizational health and strategic progress.

- **Initiative Status:** Current state of major strategic projects and programs.

- **Resource Utilization:** Financial and human resource allocation and efficiency.

- **Competitive Position:** Market share, customer satisfaction, and competitive benchmarking.

- **Risk Indicators:** Early warning signals about potential strategic risks or obstacles.

Common Dashboard Mistakes

- **Information Overload:** Including too many metrics without clear prioritization.

- **Lagging Indicator Focus:** Emphasizing historical results without adequate attention to predictive metrics.

- **Poor Visual Design:** Confusing or cluttered presentation that obscures rather than clarifies information.

- **Stale Data:** Dashboard information that is not current enough to support responsive management.

- **Lack of Context:** Metrics without sufficient background information to enable interpretation.

EVALUATE

Strategic evaluation represents the systematic assessment of strategic effectiveness and competitive impact. In the Vairos framework, evaluation goes beyond simple performance measurement to examine whether strategic initiatives are creating intended value and competitive advantage.

Evaluation vs. Measurement

- ⅄ **Measurement:** Tracking quantitative performance against established targets and benchmarks.

- ⅄ **Evaluation:** Analyzing performance information to assess strategic effectiveness, competitive impact, and organizational learning.

Evaluation adds context, interpretation, and judgment to measurement data, transforming information into strategic intelligence.

Evaluation Dimensions

- **Strategic Alignment:** Assessing whether strategic initiatives are advancing organizational mission and vision.

- **Competitive Impact:** Understanding how strategic activities affect market position and competitive advantage.

- **Resource Efficiency:** Evaluating return on strategic investments and resource utilization effectiveness.

- **Stakeholder Value:** Analyzing whether strategic initiatives create value for customers, employees, investors, and other stakeholders.

- **Organizational Learning:** Assessing what knowledge and capabilities are being developed through strategic execution.

- **Environmental Adaptation:** Understanding how well strategic initiatives respond to changing market and competitive conditions.

Evaluation Timing and Frequency

- **Quarterly Reviews:** Regular assessment of strategic progress and performance trends.

- **Annual Evaluation:** Comprehensive review of strategic effectiveness and competitive impact.

- **Project Completion:** Post-implementation evaluation of specific strategic initiatives.

- **Milestone Assessment:** Evaluation at key strategic milestones and decision points.

- **Continuous Monitoring:** Ongoing attention to evaluation indicators and trends.

Common Evaluation Mistakes

- **Metrics without Context:** Focusing on numbers without understanding their strategic significance.

- **Short-Term Bias:** Overemphasizing immediate results while ignoring long-term strategic impact.

- **Attribution Errors:** Incorrectly linking performance outcomes to specific strategic initiatives.

- **Confirmation Bias:** Seeking information that confirms existing beliefs while ignoring contradictory evidence.

- **Stakeholder Neglect:** Failing to consider impact on all affected stakeholders.

FEEDBACK

Feedback represents the critical connection between strategic execution and strategic learning, completing the Vairos planning cycle by transforming experience into insight that improves future strategic planning and execution. In the Vairos framework, feedback is not closure but continuity, simultaneously ending one planning cycle and beginning the next.

Feedback as Strategic Continuity

Feedback serves multiple essential functions that transcend simple performance review:

- **Closure and Opening:** Completing current strategic cycles while informing future planning.

- **Learning Creation:** Converting execution experience into organizational knowledge and capability.

- **Performance Improvement:** Identifying opportunities for enhanced strategic effectiveness.

- **Cultural Development:** Building organizational values that support continuous improvement and learning.

- **Institutional Memory:** Creating organizational knowledge that persists beyond individual involvement.

Feedback Sources and Methods

- **Team Retrospectives:** Structured reflections on strategic initiative outcomes and learning.

- **Leadership Debriefs:** Executive-level analysis of strategic decision quality and effectiveness.

- **Automated Dashboard Feedback:** Systematic insights from performance measurement systems.

- **Stakeholder Input:** Customer, employee, and partner perspectives on strategic impact.

- Competitive Intelligence: Market feedback about strategic positioning and effectiveness.

- **Cultural Surveys:** Organizational feedback about strategic execution and employee experience.

The Strategic Flywheel

Feedback completes the strategic flywheel that drives continuous organizational improvement:

Assess → Decide → Execute → Measure → Evaluate → Feedback → Assess

Each cycle builds on the previous, creating organizational capability that improves strategic planning and execution over time. Effective feedback ensures that strategic experience becomes strategic advantage through systematic learning and application.

Common Feedback Failures

- **Defensive Responses:** Organizational cultures that resist or punish honest feedback.

- **Information Overload:** Gathering excessive feedback without systematic processing or application.

- **Attribution Errors:** Incorrectly linking outcomes to specific causes or interventions.

- **Action Paralysis:** Collecting feedback without converting insights into organizational improvements.

- **Memory Loss:** Failing to capture and retain feedback insights for future application.

ORGANIZATIONAL STRATEGY

Organizational strategy encompasses the fundamental approaches that organizations use to create competitive advantage and achieve long-term success. In the Vairos framework, organizational strategy provides the overarching framework that guides all strategic decisions and initiatives.

Strategic Choice Framework

Organizations face a fundamental choice between two primary strategic approaches:

- **Evolutionary Strategies:** Systematic improvement and optimization of existing capabilities and market positions.

- **Revolutionary Strategies:** Transformational change that redefines organizational capabilities and competitive positioning.

This choice affects every aspect of strategic planning, from resource allocation to risk management to organizational culture.

Strategy Selection Criteria

- **Market Conditions:** Industry stability, competitive dynamics, and rate of change.

- **Organizational Readiness:** Current capabilities, resources, and change capacity.

- **Competitive Position:** Market share, competitive advantages, and strategic vulnerabilities.

- **Stakeholder Expectations:** Customer needs, investor requirements, and employee capabilities.

- **Environmental Pressures:** Regulatory changes, technological disruption, and economic conditions.

Integrated Strategic Approach

Sophisticated organizations often pursue both evolutionary and revolutionary strategies simultaneously:

- **Core Business Evolution:** Continuous improvement of existing operations and market positions.

- **New Venture Revolution:** Transformational initiatives that create new capabilities and market opportunities.

- **Portfolio Management:** Systematic coordination of different strategic approaches across organizational units.

- **Resource Allocation:** Balancing investment between optimization and transformation.

- **Risk Management:** Managing different risk profiles associated with evolutionary and revolutionary approaches.

EVOLUTIONARY STRATEGIES

Evolutionary strategies focus on the systematic improvement and optimization of existing organizational capabilities and market positions. These strategies emphasize measured progression, risk management, and sustainable competitive advantage through excellence in current operations.

When to Choose Evolutionary Strategies

Evolutionary approaches are most appropriate when:

- **Market Stability:** Operating in mature industries with predictable competitive dynamics.

- **Strong Performance:** Current organizational performance is competitive but faces pressure for improvement.

- **Optimization Opportunities:** Significant potential exists for improving existing systems, processes, and capabilities.

- **Change Consolidation:** Recent major changes require strategic consolidation rather than additional transformation.

- **Risk Sensitivity:** Organizational or stakeholder preference for measured rather than dramatic change.

Evolutionary Strategy Categories

- **Operational Excellence Strategy:** Systematic improvement of efficiency, quality, and cost-effectiveness across existing operations.

- **Customer Experience Optimization:** Enhanced engagement, satisfaction, and retention without fundamental business model changes.

- **Geographic Expansion:** Growth through entry into adjacent markets or underserved regions using existing capabilities.

- **Talent Development Strategy:** Building internal leadership, skills, and organizational capability over time.

- **Digital Maturity Strategy:** Gradual integration of technology and analytics to improve existing operations.

- **Process Innovation:** Systematic improvement of workflows, systems, and operational methods.

Evolutionary Strategy Benefits

- **Risk Management:** Lower uncertainty and more predictable outcomes compared to revolutionary change.

- **Resource Efficiency:** Building on existing investments and capabilities rather than starting over.

- **Stakeholder Confidence:** Maintaining continuity while demonstrating improvement and progress.

- **Cultural Alignment:** Working within existing organizational culture and values.

- **Competitive Sustainability:** Building advantages that are difficult for competitors to replicate quickly.

Success Factors for Evolutionary Strategy

- **Leadership Commitment:** Sustained executive support for systematic improvement initiatives.

- **Cultural Alignment:** Organizational values that support continuous improvement and excellence.

- **Systematic Approach:** Structured methodologies for identifying and implementing improvements.

- **Performance Measurement:** Clear metrics that track progress and demonstrate value.

- **Stakeholder Engagement:** Employee, customer, and partner support for improvement initiatives.

REVOLUTIONARY STRATEGIES

Revolutionary strategies involve transformational change that redefines organizational capabilities, business models, or competitive positioning. These strategies are designed to create breakthrough competitive advantages, respond to existential threats, or capitalize on major market disruptions.

When to Choose Revolutionary Strategies

Revolutionary approaches become necessary when:

- **Market Disruption:** Fundamental changes in industry structure, technology, or customer expectations.

- **Competitive Threats:** Existential challenges that cannot be addressed through incremental improvement.

- **Growth Ceilings:** Current business models have reached inherent limits that optimization cannot overcome.

- **Technology Transformation:** New technologies enable entirely different approaches to value creation.

- **Customer Evolution:** Dramatic shifts in customer needs that require new value propositions.

- **Regulatory Changes:** Policy shifts that fundamentally alter industry dynamics and competitive requirements.

Revolutionary Strategy Categories

- **New Venture Launch:** Creating entirely new business lines, services, or digital divisions.

- **Business Model Transformation:** Fundamental changes in how value is created and captured.

- **Technology Disruption:** Integration of transformational technologies that redefine operations or market position.

- **Strategic M&A:** Acquisitions or divestitures that reshape organizational portfolio and capabilities.

- **Brand Reinvention:** Comprehensive changes to organizational identity, mission, and market positioning.

- **Market Creation:** Developing entirely new market categories or customer segments.

Managing Revolutionary Risk

- **Staged Implementation:** Phased approach that tests concepts before major resource commitments.

- **Portfolio Approach:** Pursuing multiple revolutionary initiatives to diversify risk.

- **Learning Orientation:** Treating revolutionary initiatives as experiments that generate knowledge.

- **Failure Management:** Systems for recognizing unsuccessful initiatives and reallocating resources quickly.

- **Success Scaling:** Capabilities for rapidly expanding successful revolutionary initiatives.

Success Factors for Revolutionary Strategy

- **Leadership Courage:** Willingness to make bold decisions and invest in uncertain outcomes.

- **Resource Commitment:** Adequate financial and human resources for revolutionary success.

- **Cultural Flexibility:** Organizational ability to adapt to fundamental changes in operations and identity.

- **Market Timing:** Launching revolutionary initiatives when market conditions favor transformation.

- **Execution Excellence:** Strong project management and implementation capabilities for complex change.

Balancing Evolution and Revolution

- **Dual-Track Strategy:** Simultaneously optimizing current operations while developing revolutionary capabilities.

- **Resource Management:** Balancing investment between stability and transformation.

- **Cultural Integration:** Managing tension between operational excellence and revolutionary innovation.

- **Timeline Coordination:** Sequencing evolutionary and revolutionary initiatives for maximum impact.

- **Learning Transfer:** Applying insights from evolutionary improvement to revolutionary development.

CONCLUSION

The Vairos Planning System provides a comprehensive framework for data-driven strategic planning that bridges the gap between organizational capability and market opportunity. By systematically moving through the phases of Assess, Decide, and Execute, organizations can transform data into strategic advantage while building the capabilities necessary for sustained competitive success.

The power of Vairos lies not in any single component but in the integration of analytical rigor with execution discipline. Whether pursuing evolutionary improvement or revolutionary transformation, the framework provides the structure and tools necessary for strategic excellence in dynamic competitive environments.

Strategic planning is not a destination but a capability, the ability to systematically assess reality, make informed decisions, and execute with discipline while learning and adapting continuously. The Vairos framework builds this capability by providing clear methodologies, practical tools, and integrated processes that transform strategic planning from episodic activity into continuous organizational advantage.

The CEO's Role

As a CEO, your role is not to perform every analytical task or execute every tactical initiative.

Your role is to:

- ⅄ Ensure that your organization has the right framework for strategic planning and execution.

- ⅄ Build and empower teams that can assess, decide, and execute with excellence.

- ⅄ Create accountability systems that drive results and learning.

- ⅄ Model the discipline and courage that strategic success requires.

- ⅄ Maintain focus on the vital few priorities that will create competitive advantage.

The Path Forward

Victory comes to those who plan systematically, decide courageously, and execute with discipline. The Vairos Planning System provides the methodology for achieving victory in the right and opportune moment.

The framework is your guide. The execution is your responsibility. The results will be your legacy.

Begin today. Assess your current position with clarity and honesty. Decide with courage and conviction. Execute with discipline and determination. Measure, evaluate, learn, and improve. The cycle never ends, and neither does the opportunity for competitive advantage.

The future belongs to organizations that master Data Driven Strategy. With Vairos as your guide, that future can be yours.